INTRODUCTION

Viagra relaxes muscle groups found inside the walls of blood vessels and will increase blood waft to particular regions of the frame. Viagra is used to deal with erectile disorder (impotence) in guys. Another emblem of sildenafil is Revatio, which is used to treat pulmonary arterial high blood pressure and improve exercising potential in men and women. This web page contains precise statistics for Viagra, now not Revatio. Do not take Viagra while also taking Revatio, until your health practitioner tells you to.

BEFORE TAKING THIS MEDICINAL DRUG

You ought to now not use Viagra if you are allergic to sildenafil, or:

- In case you take other drug treatments to deal with pulmonary arterial high blood pressure, inclusive of riociguat (Adempas).

- If you take nitrates.

Do now not take Viagra in case you also are the usage of a nitrate drug for chest ache or heart issues. This consists of nitroglycerin, isosorbide dinitrate, and isosorbide mononitrate. Nitrates also are observed in a few leisure capsules including amyl nitrate or nitrite ("poppers"). Taking sildenafil with a nitrate medicine can reason a surprising and extreme lower in blood pressure.

To make sure Viagra is safe for you, tell your doctor when you have ever had:

- Coronary heart disease or coronary heart rhythm issues, coronary artery sickness;

- A heart assault, stroke, or congestive heart failure;

- Excessive or low blood stress;

- Liver or kidney disease;

- A blood cell disorder along with sickle cellular anemia, a couple of myeloma, or leukemia;

- A bleeding ailment which include hemophilia;

- A belly ulcer;

- Retinitis pigmentosa (an inherited situation of the eye);

- A physical deformity of the penis (inclusive of Peyronie's disorder); or

- If you have been advised you must no longer have sexual sex for fitness motives. Viagra can decrease blood waft to the optic nerve of the eye, causing unexpected imaginative and prescient loss. This has passed off in a small number of humans taking sildenafil, maximum of whom also had coronary heart disease, diabetes, excessive blood stress, excessive ldl cholesterol, or sure pre-current eye problems, and in individuals who smoked or were over 50 years antique. It isn't always clear whether or not sildenafil is the actual cause of vision loss. Viagra is not predicted to damage an unborn infant. Tell your doctor if you are pregnant or plan to become pregnant. It is not acknowledged whether sildenafil passes into breast milk or if it may harm a nursing child. Tell your doctor if you are breast-feeding a toddler. Do no longer give this medication to all of us underneath 18 years vintage without scientific recommendation.

VIAGRA ASPECT EFFECTS

Get emergency clinical assist when you have symptoms of a hypersensitivity to Viagra: hives; trouble respiratory; swelling of your face, lips, tongue, or throat. Stop taking Viagra and get emergency clinical assist when you have:

- Coronary heart attack signs and symptoms--chest ache or pressure, ache spreading for your jaw or shoulder, nausea, sweating;

- Vision modifications or surprising imaginative and prescient loss; or

- Erection is painful or lasts longer than four hours (extended erection can harm the penis). Call your doctor right away when you have:

- ringing in your ears, or unexpected listening to loss;

- Irregular heartbeat;

- swelling for your palms, ankles, or ft;

- Shortness of breath;

- Seizure (convulsions); or

- A mild-headed feeling, like you would possibly skip out. Common Viagra side results may additionally encompass:

- flushing (warmth, redness, or tingly feeling);

- Headache, dizziness;

- Extraordinary vision (blurred vision, changes in shade vision)

- Runny or stuffy nostril, nosebleeds;
- sleep problems (insomnia);
- Muscle ache, returned ache; or
- Disenchanted belly.

This isn't always an entire list of facet outcomes and others can also occur. Call your doctor for clinical advice approximately side consequences. You can also file aspect effects to FDA at 1-800-FDA-1088.

WHAT DIFFERENT DRUGS WILL HAVE AN EFFECT ON VIAGRA?

Do not take Viagra with similar medications which include avanafil (Stendra), tadalafil (Cialis) or vardenafil (Levitra). Tell your medical doctor approximately all different medicines you use for erectile disorder. Tell your doctor approximately all of your cutting-edge medicines and any you start or forestall the usage of, specially:

- Pills to treat high blood strain or a prostate sickness;

- An antibiotic - clarithromycin, erythromycin, or telithromycin;

- Antifungal medication - ketoconazole or itraconazole; or

- Medication to treat HIV/AIDS - atazanavir, indinavir, ritonavir, or saquinavir;

- Nitrates;

- Medications used to deal with pulmonary artery hypertension. This listing isn't entire. Other pills may additionally have interaction with sildenafil, which includes prescription and over the counter drug treatments, vitamins, and herbal products. Not all viable interactions are indexed in this medicinal drug guide.

HAT MAY ADDITIONALLY ENGAGE WITH THIS MEDICATION?

Do now not take this medicinal drug with any of the following:

- Cisapride

- Nitrates like amyl nitrite, isosorbide dinitrate, isosorbide mononitrate, and nitroglycerin

- Riociguat

This remedy may interact with the subsequent:

- Antiviral medicines for HIV or AIDS

- Bosentan

- Certain medicinal drugs for benign prostatic hyperplasia (BPH)

- Certain medicines for blood strain

- Certain medicinal drugs for fungal infections like ketoconazole and itraconazole

- Cimetidine

- Erythromycin

- Rifampin

This list might not describe all possible interactions. Give your fitness care company a list of all of the medicines, herbs, non-prescribed drugs, or nutritional supplements you operate. Also tell them if you smoke, drink alcohol, or use illegal tablets. Some objects may also engage with your medicine.

WHAT NEED TO I WATCH FOR WHILST THE USE OF THIS MEDICATION?

If you be aware any changes for your imaginative and prescient while taking this remedy, name your care group as quickly as viable. Stop using this remedy and call your care group right away when you have a lack of sight in one or both eyes. Contact your care crew proper away if you have an erection that lasts longer than 4 hours or if it becomes painful. This may be a sign of an extreme trouble and should be dealt with right away to save you permanent damage. If you revel in signs and symptoms of nausea, dizziness, chest ache or arm pain upon initiation of sexual pastime after taking this medicine, you should refrain from in addition hobby and make contact with your care group as soon as possible. Do now not drink alcohol to excess (examples, 5 glasses of wine or five shots of whiskey) while taking this medicine. When taken in excess, alcohol can increase your probabilities of getting a headache or getting dizzy, growing your coronary heart fee or reducing your blood stress. Using this remedy does now not defend you or your partner against HIV contamination (the virus that causes AIDS) or different sexually transmitted diseases.

WHAT FACET OUTCOMES CAN ALSO I BE AWARE FROM RECEIVING THIS MEDICATION?

Side outcomes that you should document in your care team as soon as feasible:

- Allergic reactions—pore and skin rash, itching, hives, swelling of the face, lips, tongue, or throat

- Hearing loss or ringing in ears

- Heart attack—ache or tightness inside the chest, shoulders, hands, or jaw, nausea, shortness of breath, bloodless or clammy pores and skin, feeling faint or lightheaded

- Heart rhythm modifications—rapid or abnormal heartbeat, dizziness, feeling faint or lightheaded, chest pain, trouble breathing

- Low blood pressure—dizziness, feeling faint or lightheaded, blurry vision

- New or worsening shortness of breath

- Prolonged or painful erection

- Stroke—surprising numbness or weak point of the face, arm, or leg, hassle speaking, confusion, trouble taking walks, loss of stability or coordination, dizziness, severe headache, change in vision

- Sudden imaginative and prescient loss in one or both eyes Side consequences that typically do now not require medical attention (document on your care team if they retain or are bothersome):

- Facial flushing or redness
- Headache
- Nosebleed
- Runny or stuffy nose
- Trouble dozing
- Upset stomach

This list may not describe all viable side results. Call your physician for scientific advice about side consequences. You may additionally report side consequences to FDA at 1-800-FDA-1088.

PROPER USE

Use sildenafil precisely as directed by way of your physician. Do no longer use greater of it and do not use it greater regularly than your medical doctor ordered. If an excessive amount of is used, the chance of facet results is increased. This medicine comes with an affected person records leaflet. Read and follow those commands carefully before you start using sildenafil and whenever you get a refill of your medication. Ask your health practitioner when you have any questions. You may additionally take this medication without or with meals. To use the oral liquid:

- Shake the bottle properly for as a minimum 10 seconds before each use.

- Remove the cap with the aid of pushing it down and twisting it counterclockwise. Push the plunger of the syringe, and then insert the end into the bottle even as keeping the bottle upright on a flat floor.

•   Slowly pull again the plunger till the specified dose. If you notice air bubbles inside the syringe, repeat the stairs. Remove the syringe from the bottle. Do now not press at the plunger of the syringe.

•   Place the top of the syringe into you or your baby's mouth and factor it towards the cheek. This medicinal drug usually starts offevolved to paintings for erectile disorder inside half-hour after taking it. It keeps to paintings for up to four hours, although its movement is normally less after 2 hours.

Use most effective the brand of this medicine that your medical doctor prescribed. Different manufacturers won't paintings the equal way.

PRECAUTIONS

It is vital which you tell all your docs which you take sildenafil. If you need emergency scientific care for a coronary heart hassle, it is essential that your doctor knows while you remaining took sildenafil. Do no longer use this medicinal drug in case you also are the use of a nitrate medication, frequently used to deal with angina or excessive blood pressure. Nitrate medicines consist of nitroglycerin, isosorbide, Imdur��, Nitro-Bid®, Nitro-Dur®, Nitrol® ointment, Nitrolingual® spray, Nitrostat®, and Transderm Nitro®. Some unlawful ("avenue") tablets called "poppers" (including amyl nitrate, butyl nitrate, or nitrite). Do now not use this medicine in case you additionally use riociguat (Adempas®).

If you may be taking this medication for pulmonary arterial hypertension, your doctor will need to check your development at everyday visits. This will allow your medical doctor to look if the drugs are working well and to decide in case you need to continue to take it. If you are taking sildenafil for pulmonary arterial high blood pressure, do no longer take Viagra® or different PDE5 inhibitors, which include tadalafil (Cialis®) or vardenafil (Levitra®). Viagra® additionally includes sildenafil. If you take too much sildenafil or take it together with those medicines, the chance for facet effects might be higher. Sildenafil ought to no longer be used with every other medicine or tool that causes erections.

It is crucial to tell your health practitioner approximately any heart problems you have now or may also have had within the past. This medication can purpose serious aspect consequences in patients with heart problems. If you enjoy an extended or painful erection for four hours or greater, contact your health practitioner immediately. This situation may also require prompt clinical treatment to prevent serious and permanent damage on your penis. If you revel in an unexpected lack of imaginative and prescient in one or each eyes, contact your medical doctor at once. Check along with your medical doctor right away when you have an unexpected lower in hearing or lack of listening to, which can be followed by using dizziness and ringing in the ears.

If you already use remedy for excessive blood stress (hypertension), sildenafil should make your blood strain go too low. Call your physician right away if you have blurred imaginative and prescient, confusion, dizziness, faintness, or lightheadedness while getting up from mendacity or sitting function all at once, sweating, or uncommon tiredness or weakness. This medicine does no longer guard you against sexually transmitted diseases (which include HIV or AIDS). Use shielding measures and ask your doctor if you have any questions on this. Do no longer take other drug treatments unless they were discussed with your doctor. This consists of prescription or nonprescription (over-the-counter [OTC]) drug treatments and herbal or nutrition dietary supplements.

IS SILDENAFIL RIGHT FOR ME?

Sildenafil works well with few facet consequences for the general public. Sometimes the health practitioner may be unsure whether Sildenafil is right for you and will ask a couple of greater questions to make sure. Occasionally, Sildenafil can be risky to use. You must seek advice from your physician or our group earlier than taking Sildenafil if any of the subsequent applies;

- You take nitrates for a heart condition. These are typically pills or drugs referred to as Isosorbide Mononitrate or Nicorandil (a medicinal drug very just like a nitrate) however may be called via other names

- You use poppers

- You have angina or have ever been given a spray that is going beneath your tongue to help chest pain

- You have had a heart attack or stroke inside the last six weeks

- You have a circumstance referred to as Peyronie's or your penis has a bend while erect

- You, or all of us for your own family, have records of eye sickness, which includes retinitis pigmentosa or non-arteritic anterior ischaemic optic neuropathy (NAION)

- You have liver or kidney troubles

- You have sickle-mobile sickness, multiple myeloma, or leukaemia

- You have a belly ulcer you have low blood strain

- You have out of control high blood stress Contact us for more records about any of the above, all conversations are confidential and non-judgemental. We also ask in particular about those in our online consultation, are honest and deliver as a whole lot facts as you can. Occasionally, Sildenafil won't be an awesome match for you, or every other remedy would possibly suit you better. We will tell you if that is the case at some point of your consultation.

SILDENAFIL FOR ED

Sildenafil changed into the primary powerful oral medicinal drug for ED. In 2003 it turned into joined by means of vardenafil (Levitra) and tadalafil (Cialis). Although there are a few variations among these tablets (vardenafil starts operating sooner and tadalafil works longer), they all act in precisely the identical manner to combat ED. The three drugs are so similar due to the fact they share a not unusual target, an enzyme called phosphodiesterase-5 (PDE-5). To understand why blocking off PDE-five improves sexual function, however, you have to recognise how erections expand. Normal erections require a receptive country of thoughts, adequate levels of testosterone, and healthy arteries, veins, and nerves. But in addition they require a tiny chemical messenger known as nitric oxide (NO). It serves two crucial capabilities: transmitting the impulses of arousal between nerves and relaxing the clean muscle cells in the arteries, letting them widen and admit extra blood to the penis. Nitric oxide is vital as it indicators the arterial cells to supply cyclic guanosine monophosphate (cGMP), the chemical that increases the go with the flow of blood to the penis. But the tissues of the penis also produce PDE-five, the enzyme that breaks down cGMP. In regular occasions, the penis generates enough cGMP to provide a rigid erection and enough PDE-five to give up the erection while ejaculation is entire. In many men with ED, this intricate system is out of stability, but one of the 3 oral medicines can frequently make it right. By inhibiting

PDE-5, they increase the supply of cGMP, which for plenty guys will permit erections to increase. In take a look at tubes, vardenafil and tadalafil inhibit PDE-five a bit more efficaciously than sildenafil does. But laboratory potency doesn't always are expecting scientific activity. In this example, it means that decrease doses of the more moderen pills will achieve comparable outcomes to the older drug's achievement rate. Vardenafil is marketed in 2.5-, 5-, 10-, and 20-milligram (mg) drugs; tadalafil in doses of five, 10, and 20 mg; and sildenafil, 25, 50, and a hundred mg. Each produces better results at higher doses; however the pinnacle doses also are much more likely to purpose aspect consequences. And all the drugs are extra powerful in guys with moderate ED than in folks who are significantly impaired. Doctors have had more medical enjoy with sildenafil than its newer competitors. Still, all 3 drugs seem to gain similar consequences. In huge terms, about 70% of fellows gain. The reaction is exceptional in guys with no identifiable natural cause of ED (approximately 90%), however it is much less favorable for diabetics (approximately 50%), and it is difficult to predict for men who have been treated for prostate most cancers.

OTHER OBJECTIVES: SIDE EFFECTS

If PDE-five have been observed solely in the penis, the aspect results of these pills would be restricted to that stubborn organ; in truth, an unprecedented aspect impact is priapism, painful prolonged erections that require pressing treatment. But small amounts of PDE-5 are discovered in blood vessels in other elements of the frame. In addition, PDE-five is just one of 11 enzymes within the phosphodiesterase circle of relatives. The others are concentrated in special elements of the body. While the ED pills have a miles greater affinity for PDE-5 than the alternative PDEs, they have got a few potential to inhibit those intently related enzymes. Both elements explain why the ED pills from time to time produce aspect outcomes in many parts of the body. The maximum commonplace side effects are complications and facial flushing, which occur in approximately 15% of guys. Other reactions include nasal congestion, indigestion, and lower back pain; blue-tinged imaginative and prescient is even less common. In nearly every case, those side effects are slight and brief. But new information has introduced a rare eye ailment to the list. Nonarteritic anterior ischemic optic neuropathy (NAION) is a poorly understood disease which could cause blindness. In March 2005, doctors suggested that seven guy's skilled visual impairment inside hours of the use of sildenafil. Since then, extra instances have been pronounced to the FDA associated with sildenafil, tadalafil, and vardenafil. Fortunately, the variety of instances may be very small in relation to the millions of men who have used ED tablets successfully. It's not clear that there may be a motive-and-effect dating between ED pills and NAION. At present, the priority is not superb enough to save you guys from the usage of these pills, but all guys have to use them cautiously and responsibly. The maximum critical fear approximately ED drugs is their capacity to widen arteries enough to decrease blood stress. It's rarely a trouble in wholesome guys, but it explains an important precaution that applies to all 3 medicinal drugs. Nitrates are medicinal drugs that quickly widen arteries via increasing their deliver of nitric oxide. That's how they widen partially blocked coronary arteries in sufferers with angina. But because the nitrates and ED capsules all act on nitric oxide, they do no longer blend. Men who are taking nitrates need to in no way use any of the ED tablets. This ban consists of all preparations of nitroglycerin (brief-acting, beneath-the-tongue tablets or sprays); lengthy-appearing nitrates (isosorbide dinitrate, or Isordil, Sorbitrate, and others, and isosorbide mononitrate, Imdur, ISMO, and others); nitroglycerin patches and pastes; and amyl nitrite (so-known as poppers, used for sexual stimulation through a few men). ED capsules are pretty safe for guys with strong cardiovascular ailment who do no longer take nitrates. This organization includes patients with strong angina, preceding coronary heart assaults, slight congestive coronary heart failure, nicely-controlled hypertension, and previous strokes. But guys with current heart assaults and strokes need to wait until they've recovered completely, and patients with volatile blood pressure, lively angina, or any other complex or uncommon hassle have to maintain off and get specific scientific guidance. And men who take alpha blockers (specifically terazosin, or Hytrin, and doxazosin, or Cardura) for hypertension or benign prostatic hyperplasia (BPH) have to use ED capsules (especially vardenafil or tadalafil) with incredible warning, if at all. Studies additionally suggest that sildenafil may additionally growth respiration misery in guys with severe sleep apnea and that it slows gallbladder characteristic, which might boom the danger of gallstones. Interest in Viagra is global; however scientists in Argentina have proposed the most unusual opportunity. In a 2007 paper, they record that Viagra might also assist lower jet lag. It's now not as farfetched as it appears, for the reason that drug inhibits PDE-5 within the part of the mind that controls the frame's inner clock. But before you request a prescription in your next ride, you need to recognise that the topics in the Buenos Aires experiments have been hamsters.

OTHER OBJECTIVES: THERAPEUTIC ROLES

Scientists who noticed that the ED pills can produce aspect results in many parts of the frame are asking in the event that they also can serve healing roles past male (or woman) sexuality. Because of its seniority, maximum of the research has used sildenafil, so it is not clear if the newer medicinal drugs will fill comparable roles. But for sildenafil, as a minimum, some new makes use of seem promising. Pulmonary high blood pressure. It's the only nonsexual condition that has earned FDA approval for sildenafil. It's no longer common, and sildenafil is some distance from a cure. But given that it is a critical trouble, any advantage is maximum welcome.

When we think of blood strain, we generally think of the systemic movement, of the blood pumped from the heart's left ventricle to the aorta after which to the smaller arteries that deliver oxygen-rich blood all through the body. But to pick out up crucial oxygen, blood need to first bypass from the less effective right ventricle thru the pulmonary artery to the lungs, then again to the left facet of the coronary heart. The pressures inside the pulmonary artery are lots decrease (approximately 20/10 millimeters of mercury, or mm Hg) than inside the aorta (decrease than one hundred twenty/80 mm Hg is considered wholesome). Exercise is the most not unusual reason of growing pulmonary artery stress, but the elevation is slight and subsides promptly with relaxation.

High altitudes are every other reason that can cause mountain sickness (see underneath). Far extra serious are the huge range of lung sicknesses, vascular illnesses, coronary heart disorders, and miscellaneous situations that can reason pulmonary high blood pressure. And in number one pulmonary high blood pressure, no underlying reason is evident. Pulmonary high blood pressure reasons shortness of breath, first all through exertion but sooner or later at rest if the condition progresses. A form of treatments is available, depending on the underlying problem. And the FDA has permitted sildenafil (Revatio) in a dose of 20 mg three times an afternoon for women and men with pulmonary hypertension. Clinical trials have established progressed exercise tolerance with few aspect results.

Mountain sickness. Pulmonary hypertension is a characteristic of acute mountain illness. High altitudes produce low blood-oxygen levels. In flip, low oxygen produces a narrowing of the pulmonary arteries. The heart ought to therefore paintings more difficult, lowering the ability to workout. Sildenafil widens the pulmonary arteries. To find out if it might improve exercising potential in low oxygen situations, scientists examined 14 healthful mountain climbers in a lab in Germany and once more at a Mount Everest base camp. In the lab, the volunteers breathed 10% oxygen via a mask; at the mountain, they breathed herbal air. Under both situations, a 50-mg sildenafil tablet reduced pressures inside the lungs' blood vessels and increased the most exercise potential.

It's a small take a look at, and it's too soon to say if sildenafil will assist save you or deals with acute mountain illness. Still, a take a look at of 29 climbers mentioned that tadalafil also can reduce pulmonary artery stress at a high altitude. Raynaud's phenomenon. In affected people, publicity to the bloodless triggers spasm of the small arteries that supply blood to the fingers, ft, or both. Temporarily disadvantaged of adequate blood go with the flow; the worried digits come to be faded, bloodless, and really painful. It's a common condition, affecting up to eight% of fellows and 17% of women. In the huge majority, there are not any underlying sicknesses (primary Raynaud's), and patients do well clearly by using minimizing their publicity to bloodless. But secondary Raynaud's can complicate collagen-vascular diseases or sure other conditions. It's no longer commonplace, however secondary Raynaud's may be very painful and difficult to treat. Many medicinal drugs have been used without consistent success. But a 2005 look at of sixteen patients with severe Raynaud's phenomenon that had no longer answered to different medicines suggested gain from sildenafil in a dose of 50 mg two times a day. And a 2006 have a look at of forty Raynaud's patient's stated similar benefits from vardenafil in a dose of 10 mg twice a day. Heart disorder. Sildenafil became determined by using scientists seeking out a new medicinal drug to dilate coronary arteries. It does that, but as it widens wholesome coronary arteries extra than diseased vessels, it has no longer been a hit in treating angina. But it is able to produce other benefits for cardiac sufferers. In the primary years of the Viagra era, research on sildenafil and the coronary heart become devoted to making sure the drug changed into safe for the circulate. In maximum guys with coronary heart disorder, it's miles. But numerous studies of sufferers with congestive heart failure additionally stated that the drugs improve oxygen consumption, pulmonary artery stress, and exercising capacity in those sufferers. Research suggests that sildenafil helps the coronary heart muscle relax nicely, that may help sufferers with coronary heart failure due to diastolic disorder. Another exciting asset is sildenafil's potential to shield the coronary heart from immoderate stimulation via adrenaline. Clinical trials might be had to see which patients may benefit. Stroke. It's the most speculative use for sildenafil, and its miles from clinical software. Still, it is an interesting area of recent studies. A European take a look at of 25 men who had erectile dysfunction but no other circulatory problems found that blood vessels inside the mind responded to strain better after a 50-mg dose of sildenafil. More thrilling effects had been suggested with the aid of scientists in Michigan, who produced ischemic strokes in rats by means of briefly blockading blood vessels. During their restoration, the animals were divided into 3 agencies. One acquired no remedy, any other low-dose sildenafil, and the 0.33 high-dose sildenafil. The animals who were given sildenafil recovered better, with the very best doses generating the nice outcomes.

The scientists concluded that the medicine absolutely inspired the boom of recent nerve cells inside the brain tissue subsequent to the stroke damage. It's some distance too early to realize whether or not sildenafil may additionally at some point help humans with strokes. Stay tuned.

THE END

Made in United States
Troutdale, OR
02/04/2025